Contents

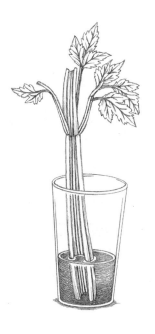

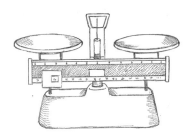

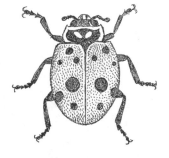

Introduction

What adjectives come to mind when you think about teaching science? If *quick* and *easy* are not on your list, *Science Question of the Day* may help put them there.

Science Question of the Day breaks down nationally required science content into easy-to-swallow, one-a-day activities. Use this collection of short, thought-provoking questions to introduce or review key topics, such as animal adaptation, ecosystems, weather, the solar system, matter, and energy. All you need is five to ten minutes a day, and by the end of the year, you'll have covered all the topics needed to satisfy the National Science Education Standards for life science, earth science, physical science, and science as inquiry.

Many of the questions in this book do double—or triple—duty. As you may well know, the No Child Left Behind Act requires all states to assess students' science knowledge as of 2007–2008 academic year. Modeled after tests currently used by several states, the problems will prepare students for the types of test questions found on science (and other) assessments. In addition, several questions give students practice in essential skills such as critical thinking, using and interpreting charts and graphs, and reading diagrams. Finally, the questions were written to tickle students' brains, provoke discussion, and possibly even motivate students into learning more about science on their own.

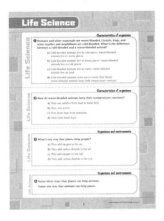

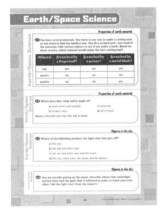

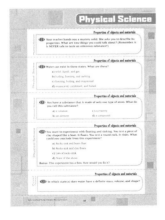

How to use this book

The book is divided into three sections—life science, earth/space science, and physical science. Each section contains 60 questions (mostly about content and a few related to inquiry skills). Each question correlates to a science standard, which is identified above the question. While the questions are numbered (so that you can match them easily to the answer key), you don't have to present the problems in any particular order. Just pick and choose questions according to the topics you're introducing or reviewing.

All the questions in this book are reproducible. You could make several photocopies of each page, cut apart the questions, and hand the same questions to students for homework or independent work. You could also make transparencies of each page and present your chosen question of the day to students, using a piece of paper to cover the other questions on the page. You might also copy individual problems on the board and challenge students to ponder over the question as they enter the classroom first thing in the morning. All the questions are perfect for jump-starting discussions on particular topics or units. Or you could hand out questions for quizzes or extra-credit exercises. There are so many possibilities! You may even find a new way of using this book!

However you choose to use these activities, try to look past the fact that they mimic science assessments, and enjoy them!

Life Science

Characteristics of organisms

❶ Humans and other mammals are warm-blooded. Lizards, frogs, and other reptiles and amphibians are cold-blooded. What is the difference between a cold-blooded and a warm-blooded animal?

a) Cold-blooded animals live in cold places; warm-blooded animals live in warm places.

b) Cold-blooded animals live in warm places; warm-blooded animals live in cold places.

c) Cold-blooded animals live in water; warm-blooded animals live on land.

d) Cold-blooded animals need sun to warm their blood; warm-blooded animals keep body temperature constant.

Characteristics of organisms

❷ How do warm-blooded animals keep their temperatures constant?

a) They use calories from food to make heat.

b) They stay active.

c) They draw heat from sunshine.

d) They burn fossil fuels.

Organisms and environments

❸ What's one way that plants help people?

a) They add oxygen to the air.

b) They add carbon dioxide to the air.

c) They add oxygen to the soil.

d) They add carbon dioxide to the soil.

Organisms and environments

❹ Name three ways that plants can help animals.

Name one way that animals can help plants.

❺ Which diagram shows a correct life cycle?

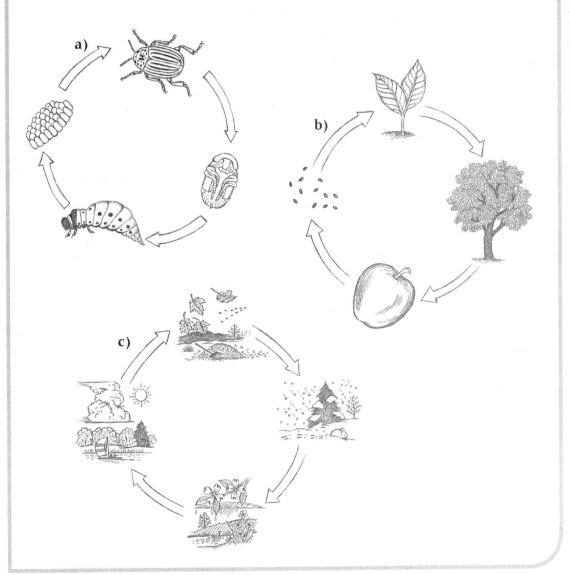

❻ Which action directly helps a plant make new seeds?

      a) Rain falls on the plant.

      b) You trim the leaves.

      c) Animals eat the flowers.

      d) Bees drink nectar from the flowers.

Bonus: Explain your answer.

Life Science

Structure and function in living systems

❼ Like all living things, trees need water. Where does a tree get most of its water?

    a) Pools of rainwater collect on the leaves.

    b) The trunk soaks up rainwater.

    c) Streams of water run down branches.

    d) Roots soak up water from the soil.

Structure and function in living systems

❽ Plants can have many different sorts of roots. What are two functions that nearly all roots share?

Structure and function in living systems

❾ A scientist is looking at tissue under a microscope. She looks at these two samples. Based on the structure of the samples, what would you infer?

 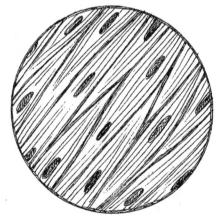

    a) They look like they perform the same function; they are both bone.

    b) They look like they perform the same function; they are both muscle.

    c) They look like they perform different functions; one is muscle and one is bone.

    d) It is impossible to infer anything.

Bonus: Identify the samples. Explain your answer.

10 What is the lowest number of cells a living organism can have?

    a) $^1/_2$ **c)** 7

    b) 1   **d)** 566,892

11 Which of the following makes its own food using energy from the sun?

    a) green algae

    b) trees

    c) sunflowers

    d) all of the above

Bonus: What do you call the process that plants use to make their own food?

12 What does a plant need to make its food?

    a) sugar, sunlight, and dirt

    b) sugar, water, and carbon dioxide

    c) sunlight, water, and carbon dioxide

    d) sunlight, water, and dirt

13 Which insect has an adaptation for drinking the nectar out of the center of flowers? (Hint: The nectar could be deep within the flower.)

a)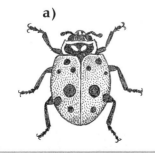
    b)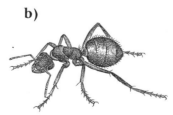
    c)

Life Science

14 **If an apple tree did not make any flowers, what would happen?**

    a) It would make seeds but not apples.

    b) It would not make any apples or seeds.

    c) It would make smaller apples than usual.

    d) It would make more apples than usual.

Life Science

15 **Groups of African elephants can journey hundreds of miles in search of food and water. Sometimes, the only water lies underground. The elephants use their long tusks to dig for water. These long tusks are an**

    a) experiment.   **c)** adaptation.

    b) exploration.   **d)** acceptation.

Life Science

16 **Different cells can do different jobs. What is one job a cell can't do?**

    a) move

    b) break down food for energy

    c) produce chemicals

    d) inhale

Life Science

17 **Which of these is not happening as your body grows?**

    a) Two cells reproduce to make a third cell.

    b) One cell divides into two.

    c) A cell grows bigger.

    d) A cell changes to take on a specific job.

Structure and function in living systems

18 What is one way that the respiratory and circulatory systems work together?

a) Blood carries oxygen to the lungs.

b) Blood carries oxygen from the lungs.

c) The lungs push blood around the body.

d) The lungs pull blood toward them.

Structure and function in living systems

19 Which organs help you digest food?

a) mouth, stomach, small intestine, large intestine

b) mouth, stomach, appendix, kidney

c) stomach, small intestine, heart

d) stomach, small intestine, large intestine, lungs

Structure and function in living systems

20 Which statement is true for both plant and animal cells?

a) Cells break down sugar for energy.

b) Cells make food from sunlight.

c) Cells use energy from food that was digested in the stomach.

d) Cells carry information through a nervous system.

Structure and function in living systems

21 Which series is written in the correct order?

a) molecules cells tissues organs

b) tissues cells molecules organs

c) cells molecules organs tissues

d) cells molecules tissues organs

22 Is it possible for a child to be exactly like its mother?

    a) No, because it doesn't get genetic material from her.

    b) No, because it gets only part of its genetic material from her.

    c) Yes, because it gets all of its genetic material from her.

    d) Yes, because it gets half of its genetic material from her.

23 Fill in this chart: Think of one inherited trait you have, and one learned trait. Add them to the chart in the correct row, as shown in the example.

Trait	Inherited	Learned
brown eyes	X	
	X	
		X

24 You put your hand under the tap to see if the bathwater is warm enough. Which organ system is most involved with helping you decide?

    a) respiratory system

    b) nervous system

    c) excretory system

    d) endocrine system

25 Imagine you're walking down the street and you see a tiger heading your way. Immediately, your body gets ready to react and run away. Name a change that will happen in one of your organ systems that could help you.

Life Science

Regulation and behavior

26 Two chimpanzees in a zoo saw bananas hanging high overhead. They pushed some boxes under the bananas and stood on the boxes to reach the food. The way they used these tools to get their food was

a) a learned trait.

b) a mistake.

c) a reflex.

d) all of the above.

Life Science

Regulation and behavior

27 If a rabbit spots a cougar, the rabbit will turn and run away. Which of its body systems are involved in its escape?

a) nervous system

b) muscular system

c) circulatory system

d) all of the above

Bonus: Besides running away, how else might an animal respond to a threat from a predator?

Life Science

Regulation and behavior

28 Animals can get bored in zoos. So some zookeepers create interesting habitats, where animals can act as they would in the wild. Read the facts about Western gorillas in the chart below. Then describe one way you could keep them interested in their zoo habitat.

Western Gorilla Traits	
diet	plants: leaves, seeds, fruits, tree bark, shoots, and flowers
activity	spend the day looking for food in the forest
social organization	live in groups of about five gorillas with one male leader
senses	similar to human senses
hands and feet	they can use hands and feet equally well

Life Science

29 Which of the following is a population?

a) All the salamanders in the world

b) All the Eastern three-toed salamanders in the world

c) All the salamanders that live in one field

d) All the Eastern three-toed salamanders that live in one field

Bonus: Which of these animals is a species?

30 On a hike through the **Big Green Forest**, you see rocks, soil, trees, smaller plants and flowers, birds, worms, insects, and mice. **What do you call all the living and nonliving things in this forest?**

a) a species   c) a food chain

b) a habitat   d) an ecosystem

31 Prairie dogs are small furry animals that live underground. To many people, prairie dogs look cute. To some snakes, owls, and ferrets, prairie dogs look like a meal. **To owls, prairie dogs are**

a) predators.   c) producers.

b) prey.   d) decomposers.

Bonus: What would owls be to prairie dogs?

32 **What do you think will happen to a forest ecosystem after a forest fire?**

a) Nothing will grow.

b) Only moss will grow.

c) Trees will grow first, then smaller plants.

d) Small plants will grow first, then trees.

Life Science

33 Answer the two questions about the food chain shown below:

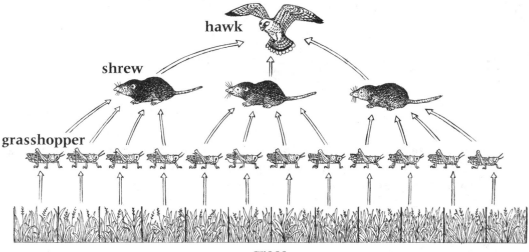

hawk

shrew

grasshopper

grass

If the population of grasshoppers suddenly decreases, which species would increase at first?

What would happen if the grasshopper population increased?

Life Science

34 A habitat gives a species what it needs to survive. For example, a forest gives a bird shelter from bad weather. Name two other ways that a habitat helps a species.

Life Science

35 Beavers change their habitat by cutting trees to build dams. The dams are their homes. What could happen to an ecosystem with too many beavers?

Life Science

36 Which type of organism adds the most nutrients to the soil as it feeds?

a) producers   c) decomposers

b) consumers   d) none of the above

37 In the ocean food chain, which of the following is a producer?

      a) seaweed   **c)** tuna fish

      b) seagulls   **d)** clams

Bonus: Name a producer in a food chain on land.

Diversity and adaptations of organisms

38 Imagine a dry and sandy habitat that gets only 4 inches of rainfall a year. Now imagine a creature that could survive there. What is one adaptation that could help this creature survive?

Diversity and adaptations of organisms

39 The desert horned lizard lives in the desert and eats ants and other small animals. Its predators include foxes, coyotes, and eagles. Fill in the chart to explain how each of the lizard's adaptations helps it survive.

Adaptation	How it helps the lizard survive
black-brown skin blends in with desert sand and rocks	
good sense of smell	
fast reflexes	
thick skin (water does not pass easily through it)	

Life Science

Diversity and adaptations of organisms

40 **Which event could cause a species to go extinct?**

    a) a fallen meteorite covering much of earth in dust and blocking out the sun

    b) a change in climate

    c) the loss of habitat

    d) all of the above

Bonus: Explain how one of the above could lead to extinction.

Diversity and adaptations of organisms

41 **Which paragraph correctly describes how species of fish adapt to changes in their environment?**

    a) In every generation, fish are born with slightly different traits. Some traits help the fish survive the changing environment. Fish with these helpful traits survive and pass on those traits.

    b) Not all fish that are born can survive in the changing environment. In their lifetimes, some of these fish can change to better fit the environment. These fish pass on the new traits to their offspring.

Personal health

42 **Which action could most directly result in you catching a cold?**

    a) standing in the rain

    b) getting cold feet

    c) someone sneezing on you

    d) eating the wrong food

Personal health

43 **Which statement is correct?**

    a) Bacteria cause all diseases.

    b) Viruses cause all diseases.

    c) Bacteria and viruses cause many diseases.

    d) Bacteria and viruses don't cause diseases.

44 Pretend that you have a friend who catches a lot of colds. Give this friend two science-based tips for catching fewer colds.

45 Your friend has an unhealthy diet full of sodas, candy, and fast food. What advice might you give?

   a) Eat fewer nutrients and more carbohydrates.

   b) Eat less fat, less sugar, and fewer nutrients.

   c) Eat less sugar and more protein and minerals.

   d) Eat fewer calories and more fat.

Bonus: In addition to eating right, name two things that are important for healthy living.

46 Which is a bad thing to have in your food, even if taken in the right amounts?

   a) protein   **c)** calories

   b) minerals   **d)** none of the above

Bonus: Plan a menu. Which foods would you prepare to have a healthy dinner?

47 Use this list of activities to answer the questions below. Each question has more than one answer.

   1. Washing their hands   **4.** Eating right

   2. Brushing their teeth   **5.** Avoiding sugar and fat

   3. Getting enough sleep   **6.** Exercising

Which of these helps keep people from getting sick?

Which of these helps keep people from becoming overweight?

Life Science

Abilities necessary to do scientific inquiry

48 You want to find the mass of an animal. Which unit could you measure it in?

    a) degrees   **c)** kilometers

    b) square inches   **d)** kilograms

Abilities necessary to do scientific inquiry

49 To learn more about elephant migration, what should scientists do?

    a) Bring some of the elephants to a zoo so they can observe them closely.

    b) Create obstacles and observe how elephants travel around them.

    c) Put out some food and observe how elephants move toward it.

    d) Try to observe the elephants without changing their habitat.

Abilities necessary to do scientific inquiry

50 After you do an experiment, what's one way to communicate your results?

    a) make a chart

    b) make a graph

    c) write a paragraph to explain your conclusion

    d) all of the above

Abilities necessary to do scientific inquiry

51 You are designing an experiment to find out which type of frog can jump farther. Which is a controlled variable (variable that doesn't change) in this experiment?

    a) jumping distance

    b) type of frog

    c) color of the frog

    d) texture of the jumping surface

Bonus: Name one tool that you would need for the above experiment.

52 A scientist wants to compare two cells: one from a liver and one from a heart. Which tool would be most helpful?

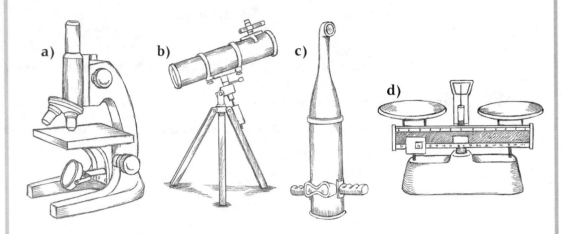

a)   b)   c)   d)

Bonus: Name a job that one of the other tools could do.

53 A scientist wants to find out which toy car rolls the fastest. She rolls each car down a slope as shown. What is wrong with this experiment?

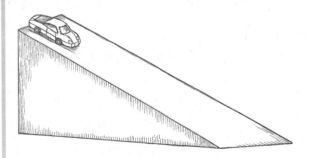
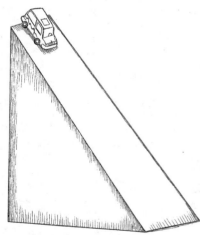

a) There are not enough variables.

b) There are too many variables changing.

c) There are not enough cars.

d) The slopes should be steeper.

54 You conducted an experiment to find the best conditions to grow a flower called a daylily. Here are the results. Based on the chart, under which condition does a daylily grow best?

Conditions for plant growth	Plant height
full sun	7 cm
partial sun	9 cm
full shade	4 cm

a) full sun

b) partial sun

c) full shade

d) can't tell from the chart

55 You set up an experiment to find out how plants grow in crowded conditions. You set up two cartons with soil, seeds, and water. What is the one variable that you should change?

a) type of soil

b) number of seeds

c) amount of water

d) amount of sunlight

56 On a walk through the park, you observe squirrels and chipmunks. You would like to learn more about them. Which of these questions could best be tested scientifically?

a) Are squirrels in this park afraid of people?

b) Are squirrels more afraid of people than chipmunks are?

c) Are chipmunks very cute?

d) Are chipmunks in this park cuter than the squirrels?

Bonus: For the question you chose above, write a hypothesis that you could test with an experiment.

Life Science

57 The data on this graph show how the height of bean plants changes when the plants grow crowded together. If you planted 30 bean plants per square meter, about how high would you expect the plants to grow?

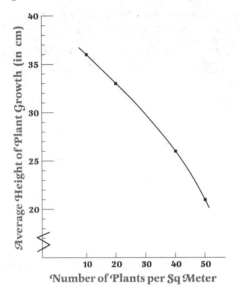

a) 44 cm

b) 36 cm

c) 30 cm

d) 18 cm

Bonus: What conclusion would you draw from this experiment?

58 Which thermometer shows the ideal daytime temperature for a plant called a gardenia?

Ideal temperature for gardenias (degrees Fahrenheit)	
daytime	65 – 70
nighttime	60 – 62

a)

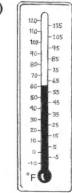

b)

c)

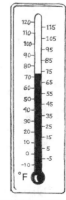

Life Science

59 Your teacher shows you this demonstration: She puts a piece of celery in water. She adds blue food coloring to the water. The celery pulls the water up its stalk. This diagram shows you the results. What is your teacher demonstrating?

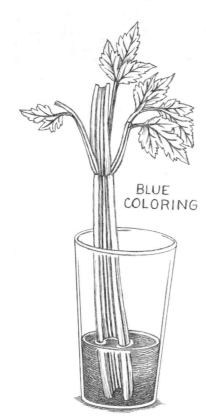

BLUE COLORING

a) How food coloring spreads

b) How leaves turn blue

c) How plants pull water to their leaves

d) How blue and green paint can mix

Life Science

60 Your teacher challenges you to design an experiment that needed a balance. Which of these would do it?

a) Compare the buoyancy of two types of clay

b) Compare the sliding speed of two snails

c) Compare the friction created by two rolling toy cars

d) Compare the mass of two types of rock

Earth/Space Science

Properties of earth material

61 You have several minerals. You want to use one to make a cutting tool, so you want to find the hardest one. You do a scratch test—you scratch the minerals with various objects to see if you make a mark. Based on these results, which mineral would make the best cutting tool?

Mineral	Scratched by a fingernail?	Scratched by a penny?	Scratched by a metal blade?
talc	yes	yes	yes
quartz	no	no	no
apatite	no	no	yes

Properties of earth material

62 Which describes what soil is made of?

    a) dead plants and animals   **c)** minerals

    b) broken rocks   **d)** all of these

Bonus: Describe one way that soil is made.

Objects in the sky

63 Which of the following produce the light that they give off?

    a) the sun

    b) the sun and other stars

    c) the sun and other stars and the moon

    d) the sun, other stars, the moon, and the planets

Objects in the sky

64 You are outside gazing at the moon. Describe where that moonlight started from and the path that it followed in order to reach your eyes. (Hint: Did the light start from the moon?)

Earth/Space Science

65 Pictures of the moon show many craters. What caused most of these?

    a) mysterious explosions

    b) "moon quakes"

    c) space rocks hitting the moon

    d) scientists don't know

Earth/Space Science

66 It is a clear, dark night. You have a good view of the sky. You see plenty of stars, but you cannot see the moon. What could explain this?

    a) Light from the stars makes it hard to see the moon.

    b) There is a star or planet blocking the moon.

    c) The side of the moon that's lit up is facing away from the Earth.

    d) Earth is blocking the sunlight that ordinarily reaches the moon.

Earth/Space Science

67 This is a diagram of a solar eclipse. A solar eclipse happens when

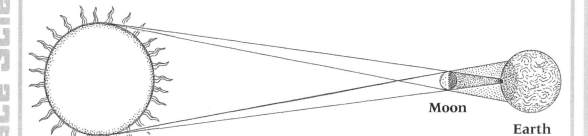

Sun   Moon   Earth

a) Earth casts a shadow on the sun.   **c)** Earth casts a shadow on the moon.

b) the sun casts a shadow on Earth.   **d)** the moon casts a shadow on Earth.

Earth/Space Science

68 Which action causes day to turn to night?

    a) The moon casts a shadow.

    b) The moon orbits Earth.

    c) Earth orbits the sun.

    d) Earth rotates on its axis.

Earth/Space Science

69 How do you know where to find the sunset each night?

    a) Look to the right.

    b) Look to the east.

    c) Look to the west.

    d) It depends where you are.

Bonus: How do you find the sunrise?

Earth/Space Science

70 Seasons change as

    a) Earth rotates on its axis.

    b) Earth revolves around the sun.

    c) Earth changes shape.

    d) the moon revolves around Earth.

Earth/Space Science

71 Seasons change because

    a) Earth's orbit is not round.

    b) Earth's axis is tilted.

    c) Earth's orbit changes speed.

    d) Earth's revolution around its axis slows down.

72 Water is a solvent. This property of water results in which of the following during the water cycle?

    a) Rain has bits of rock in it.

    b) Hail is hard.

    c) Rivers stop running.

    d) Minerals are washed into the sea.

73 When water mixes with rock, tiny particles of the rock dissolve in the water. The rock-and-water mixture is called a

    a) solvent.

    b) solution.

    c) dissolution.

    d) dissolver.

74 Wind would not blow if which thing didn't happen?

    a) Trees didn't shake.

    b) The ocean didn't have waves.

    c) The sun didn't heat the air.

    d) Earth didn't revolve around the sun.

75 Which would make the best model to show the structure of the inside of the Earth?

    a) a basketball   **c)** a hard-boiled egg

    b) solid rock   **d)** a rubber-band ball

Bonus: Explain why your model is a good one.

Earth/Space Science

Earth/Space Science

76 Imagine you had a machine that could dig straight through Earth to the other side. Which problem would it have to overcome?

a) intense heat

c) fire

b) intense cold

d) strong acid

Earth/Space Science

77 Rain is pouring down on you. Your friend asks where so much water came from. You explain that the water could once have been

a) in a lake.

c) in an ocean.

b) in a river.

d) any of the above.

Earth/Space Science

78 Which could make the best model of the Earth's water cycle?

a) a tank full of soil, with a cover

b) a tank full of soil and water, with a cover

c) a tank full of soil and water, uncovered

d) a tank full of water, uncovered

Earth/Space Science

79 Which of these add gases to Earth's atmosphere?

a) cars

c) people

b) plants

d) all of the above

Bonus: Which of these use the atmosphere's gases?

Earth/Space Science

80 Which process does the water cycle depend on?

a) combustion

c) evaporation

b) elongation

d) dehydration

Science Question of the Day Scholastic Teaching Resources

81 Earth is surrounded by a layer of gas. What is this layer called?

    a) the biosphere

    b) the atmosphere

    c) the hydrosphere

    d) the mantle

Bonus: Describe one way in which the layer of gas above affects the Earth.

82 Which diagram shows the path of Earth's water in the water cycle?

a)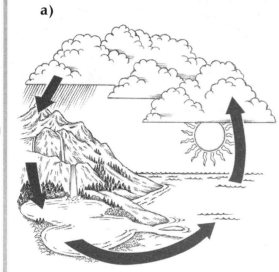

b)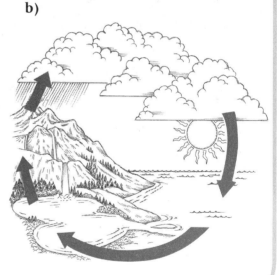

83 The sun shines in space as it does on Earth. Yet it is freezing cold in space. What causes this difference?

    a) Earth has trees and space does not.

    b) Earth has air (atmosphere) and space does not.

    c) Earth is round and space has no shape.

    d) Earth has a molted (melted) core.

Bonus: Explain your choice.

Structure of the earth system

84 Make a prediction: The pilot of a small plane parachutes out when the engine stalls. Is she more likely to land on water or on dry land?

Bonus: What fact(s) did you use to make your prediction?

Structure of the earth system

85 You are hiking on a rocky trail. You start thinking about the history of those rocks. You know that rocks could once have been

    a) soil in a field.

    b) melted rock under the Earth.

    c) lava in a volcano.

    d) all of the above.

Bonus: Describe one way in which rocks form.

Structure of the earth system

86 Very slowly, parts of Earth's crust are moving. What allows them to move?

    a) They stretch like rubber bands.

    b) They are soft like chewing gum.

    c) They are floating on melted rock.

    d) They are floating on pools of water.

Structure of the earth system

87 Tree roots can push through rock and break it. What is one other way that rock wears away?

Structure of the earth system

88 Red-hot lava (melted rock) is pouring out of an erupting volcano. What provided the heat to melt the rock?

    a) the sun   **c)** the inside of the Earth

    b) another volcano   **d)** a chemical reaction

89 Here is a map showing the location of volcanoes in the Pacific Ocean. What could explain the pattern?

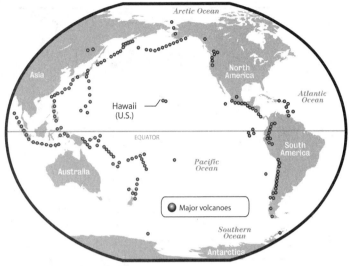

a) Volcanoes create new volcanoes.

b) Volcanoes occur where tectonic plates come together.

c) Volcanoes occur where water is the deepest.

d) Volcanoes occur where there is less mud on the ocean floor.

90 Earth's crust is made of giant pieces of rock called tectonic plates. When two of these plates push against each other, what might happen?

a) trees will die

b) mountains will slowly grow

c) oceans will form

d) people will hear a crash

91 Which effect does the ocean *not* have on the land?

a) collecting salt from runoff water

b) providing water that rains back onto land

c) eroding the shoreline

d) cooling land in winter

92 Based on this diagram, where is an earthquake most likely to occur? Circle your answer.

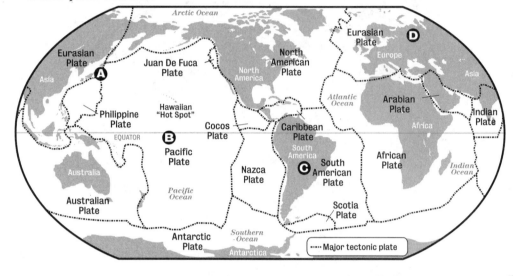

93 What will happen when it rains more than expected in an area?

    a) Lakes can get bigger.   **c)** More water can flow to the sea.

    b) The land can get flooded.   **d)** All of the above.

94 Earth's water is found in many places. Where would you find the greatest quantity?

    a) in rivers and lakes   **c)** in the air

    b) in groundwater   **d)** in the oceans

Bonus: In which of the above choices would you find the greatest amount of water vapor?

95 At the bottom of the deepest ocean, what is the ocean floor made of?

    a) rock and soil   **c)** lava

    b) ice   **d)** none of the above

Structure of the earth system

96 Which of these can lessen the erosion of a hillside?

a) trees

c) streams

b) hikers

d) power lines

Bonus: Name one thing that can trigger a landslide.

Earth's history

97 Scientists found a fossil of tooth. It came from a large, now-extinct animal called a mammoth. The tooth had a large flat surface. Based on this, what was the mammoth's diet likely to have been?

 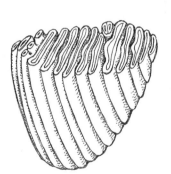

a) meat

c) fish

b) grass and leaves

d) no way to tell

Earth's history

98 After digging 30 meters underground, scientists find a rock that looks like this. Which would be a logical conclusion?

a) It is probably an oddly shaped rock.

b) It is the imprint of an animal that died very recently.

c) It is the imprint of an animal that died long ago.

d) It is a recent carving made to look like an animal.

99 Pretend you had a time machine and went back one million years. What do you think Earth's mountain ranges would look like?

a) All mountain ranges would be bigger than they are now.

b) All mountain ranges would be smaller than they are now.

c) Some mountain ranges would be bigger than now, some smaller.

d) They would all be the same size as they are now.

Bonus: What might Earth's mountains look like if you went *forward* one million years?

100 High in the mountains, scientists find a fossil of a sea animal. Which would be a logical conclusion?

a) Before there were mountains there, there was an ocean.

b) Sea animals once lived in the mountains.

c) Rain carried the animal a long distance.

d) Rain carried the fossil a long distance.

101 Which fossil is likely to be the oldest? Circle your answer.

Earth/Space Science

102 If you want to explain what Earth's moon is made of, which of these would you say it is most like?

    a) Earth   **c)** the sun

    b) Jupiter   **d)** a comet

Bonus: Choose one planet in the solar system. Describe one way that our moon is like that planet and one way that it is different.

Earth/Space Science

103 Imagine that you had a super-rocket that could take you to Mars in one month. About how long would it take you to reach the sun?

    a) one day   **c)** two months

    b) one week   **d)** ten years

Bonus: Discuss how long it might take the rocket to reach the nearest star, after the sun.

Earth/Space Science

104 How are the eight planets in our solar system similar?

    a) They are made of rock.

    b) They are the same size.

    c) They are round.

    d) All of the above

Earth/Space Science

105 Based on Jupiter's position in the solar system, how long is a year on Jupiter?

    a) longer than 365 days

    b) shorter than 365 days

    c) 365 days

    d) position does not affect the length of a year

Earth/Space Science

106 How does the sun's gravity affect Earth?

a) It speeds up Earth's rotation.

b) It heats up Earth.

c) It creates ocean tides.

d) It holds Earth in orbit.

Bonus: Which answer explains how the moon's gravity affects Earth?

Earth/Space Science

107 A planet closer to the sun than Earth is likely to be which of the following?

a) full of living things

b) made of gas

c) hotter than Earth

d) all of the above

Earth/Space Science

108 This bar graph shows the size (diameter) of the planets in the solar system. Two bars are missing labels. Figure out which two planets are missing, and label the correct bars.

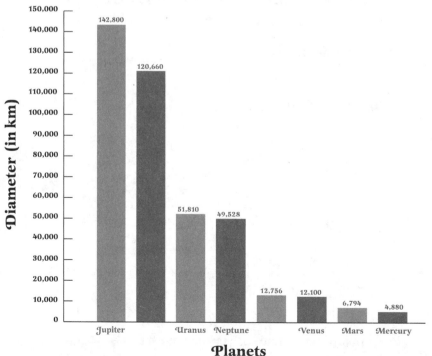

Jupiter 142,800   120,660   Uranus 51,810   Neptune 49,528   12,756   Venus 12,100   Mars 6,794   Mercury 4,880

Diameter (in km)

Planets

Science Question of the Day Scholastic Teaching Resources

Earth in the solar system

109 If the size of the sun were represented by a soccer ball, which would best represent Earth?

a) a basketball

c) a giant pumpkin

b) a bowling ball

d) a marble

Earth in the solar system

110 Complete the chart to compare Earth to Jupiter.

	Jupiter	Earth
Time it takes to orbit the sun	4,333 days	
Number of moons	63	
Length of a day	10 hours	
Size (diameter)	143,000 km	Circle one: 210,000 km 12,800 km 1,280 km

Abilities necessary to do scientific inquiry

111 Which tool would be useful for measuring different kinds of weather?

a) thermometer   b) barometer   c) meterstick   d) all of the above

112 Scientists who want to study volcanoes can use which of these tools?

   a) seismometers that measure vibrations in the ground

   b) gas meters that sample gases in the air

   c) telescopes that see smoke coming out of the top

   d) all of the above

113 You want to do an experiment to find out which kind of soil holds the most water: clay, sand, or garden soil. Before you begin your experiment, what should you do?

   a) Observe the soils and form a hypothesis.

   c) Draw a conclusion.

   b) Find a fourth type of soil to test.

   d) Wet all the soils.

Bonus: Name one variable you will keep the same and one variable you will change in this experiment.

114 Your friend designs an experiment to answer this question: *Which heats up faster—soil, air, or water?* She sets up the experiment as shown. You find a problem with her procedure. What does she need to change?

   a) She needs two containers each of soil, water, and air.

   b) She needs to mix the soil, water, and air.

   c) She needs equal volumes of soil, water, and air.

   d) She needs different types of soil.

Science Question of the Day Scholastic Teaching Resources

115 Students wanted to test the properties of three different types of soil. They wet each soil, packed it into a bucket, and turned the bucket over. Here are the three soils after the test. What would you conclude about a property of these soils?

Soil A   Soil B   Soil C

Bonus: Describe a test that you could use to learn about a different property.

116 Which tool would be best for studying Mars?

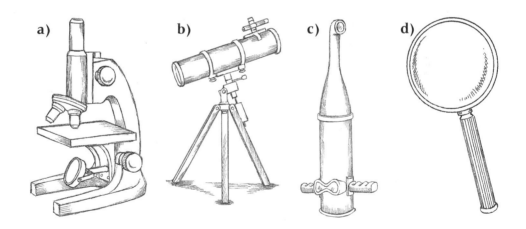

a)   b)   c)   d)

Bonus: Which tools would be best for studying samples of moon rocks that astronauts brought back?

117 Circle the tools you might use to test the properties of a rock.

microscope   piece of paper   ruler

knife   beaker of water

118 You set up this experiment: You have two equal-sized pots with two identical plants. Each plant was watered one week ago. You pour three cups of water into each pot. In pot A, no water comes out the bottom. In pot B, two cups of water come out the bottom. What could you conclude from the experiment?

a) Plant A needs more water than plant B.

b) Pot B absorbs more water than pot A.

c) Soil A holds more water than soil B.

d) Water acts differently under different conditions.

119 You set up an experiment to find out how different soils might react in an earthquake. You make two containers of soil as shown. For your experiment, you will shake container B and drop container A from the height of your desk. What is a problem with your procedure?

a) Soil does not react to earthquakes.

c) There are not enough soil types.

b) The conclusions are wrong.

d) Too many variables are changing.

120 How do you think scientists know what the deep layers of Earth are made of?

a) They compare it to the inside of other planets.

b) They dig deep holes and look.

c) They use radar and other tools to make a picture.

d) They guess.

Physical Science

Properties of objects and materials

121 Your teacher hands you a mystery solid. She asks you to describe its properties. What are two things you could talk about? (Remember, it is NEVER safe to taste an unknown substance!)

Properties of objects and materials

122 Water can exist in three states. What are these?

    a) solid, liquid, and gas

    b) boiling, freezing, and melting

    c) freezing, boiling, and evaporated

    d) evaporated, condensed, and boiled

Properties of objects and materials

123 You have a substance that is made of only one type of atom. What do you call this substance?

    a) a solution   **c)** a property

    b) an element   **d)** a compound

Properties of objects and materials

124 You want to experiment with floating and sinking. You test a piece of clay shaped like a boat. It floats. You test a round rock. It sinks. What could you conclude from this experiment?

    a) Rocks sink and boats float.

    b) Rocks sink and clay floats.

    c) Lots of rocks sink.

    d) None of the above.

Bonus: This experiment has a flaw. How would you fix it?

Properties of objects and materials

125 In which state(s) does water have a definite mass, volume, and shape?

Science Question of the Day Scholastic Teaching Resources   41

Physical Science

126 Your science teacher comes in with many liquids. She slowly pours them into a glass container, one by one. Instead of mixing, they form layers like this drawing shows. What about the liquids keeps them separate?

a) Some liquids have greater density.

b) Some liquids are heavier.

c) Some liquids are more magnetic.

d) Some liquids have greater volume.

Bonus: Circle the liquid that has the lowest density.

Physical Science

127 What is a molecule?

    a) a combination of compounds

    b) a combination of atoms

    c) the building blocks of an atom

    d) the building blocks of a neutron

Physical Science

128 It's cold outside, and you don't have any gloves. You ask your aunt the scientist for something to wrap your hands in. Which kind of material do you ask for?

    a) a material with a low boiling point

    b) a material with high mass

    c) a material that is a good conductor

    d) a material that is a good insulator

129 Scientists take a sample of a tree, a rock, and a piece of plastic. They break each down into the elements that create them. What would you expect to see?

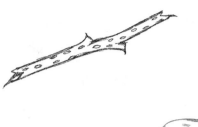
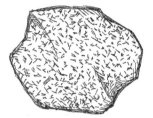

a) The three samples share some of the same elements.

b) All three samples share exactly the same elements.

c) The three samples are made of completely different elements.

d) The three samples are not made of elements.

130 Read the chart, then choose the element below that is most likely to be a metal. Fill in your choice here:

Sample _____ is more likely to be a metal.

Properties of Metals	Properties of Sample X	Properties of Sample Y
good conductors of heat	excellent conductor of heat	good conductor of heat
good conductors of electricity	does not conduct electricity	good conductor of electricity

Bonus: You receive a sample of a mystery metal Z. Write a hypothesis about this metal's ability to conduct electricity.

Properties of objects and materials

131 Which of the following is too small to be seen with an ordinary microscope?

a) an atom

c) a neuron

b) a cell

d) muscle tissue

Position and motion of objects

132 From this graph, what is occurring with the object's motion?

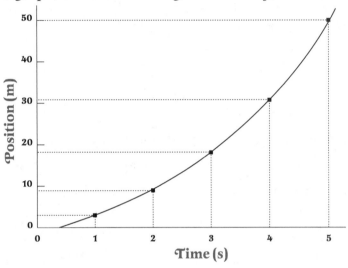

a) It is impossible to tell.

c) It is speeding up.

b) It is slowing down.

d) It is remaining the same.

Position and motion of objects

133 Susan needed extra power to open this wooden crate. So she's using a tool called a crowbar. Which of these simple machines is Susan using?

a) a pulley

b) an inclined plane

c) a wheel

d) a lever

134 There is an object on a table. Light travels from the image to your eyes. You see the image clearly because the following enters your eyes:

a) refracted light

c) absorbed light

b) transmitted light

d) reflected light

135 Which diagram correctly shows the path that allows you to see the cat? Circle your answer.

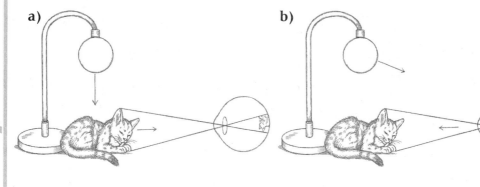

a)

b)

136 As it moves through glass, light can do which of the following?

a) refract

c) release

b) reflect

d) a and b

137 When a beam of light is broken up into its parts, what do you see?

a) darkness

c) a rainbow

b) red color

d) a flashing light

138 In which of these diagrams will the lightbulb light?

a)

b)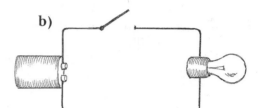

c)

139 Which of these explains how microscopes and telescopes make objects appear larger?

    a) Heat is produced by light.

    b) Lenses make light hotter.

    c) Lenses change the path light travels.

    d) Rays bounce off of objects.

140 What form does the sun's energy have when it reaches Earth?

    a) visible light

    b) infrared energy (heat)

    c) ultraviolet radiation

    d) all of the above

Light, heat, electricity, and magnetism

141 If you were creating an electrical circuit and ran out of wire, which of these materials would make a good substitute?

a) yarn

c) plastic knitting needle

b) licorice strings

d) tin foil

Bonus: Think about your choice above. What property of this material made it a good choice?

Light, heat, electricity, and magnetism

142 If you hold two magnets together, what will happen?

a) They will attract each other.

b) They will repel each other.

c) Either a or b.

d) Neither a nor b.

Light, heat, electricity, and magnetism

143 Which tool relies on the fact that the Earth has a magnetic field, with north and south poles?

a) a barometer

b) an electron microscope

c) a compass

d) a rocket

Properties and changes of properties in matter

144 Your teacher mixes together two clear liquids. Together, the two liquids bubble and turn blue. You just witnessed a

a) transfer of energy.

b) physical reaction.

c) chemical reaction.

d) force.

145 Your teacher sets out two shallow cups. She fills one with freshwater and one with saltwater. Two days later, the water has evaporated and you're left with the setup as shown.

What could you conclude?

    a) Saltwater does not evaporate.

    b) Saltwater evaporates completely.

    c) Salt from saltwater doesn't evaporate.

    d) Water doesn't evaporate.

146 You put out two glasses of water, one warm and one ice-cold. After 30 minutes, you check them and find that the cold glass has beads of water coating the outside. Where did that water come from?

    a) the air   **c)** the ice

    b) inside the glass   **d)** the water

Bonus: Explain what made the water drops form on the glass.

147 You can't decide what to have for breakfast. So you put four kinds of cereal in a bowl, add some banana slices, and stir it up. What would you call the contents of the bowl?

    a) a compound   **c)** a solution

    b) an element   **d)** a mixture

Properties and changes of properties in matter

148 You take a ball of clay and shape it into a horse. Name one physical property of the clay that changed and one that stayed the same.

Properties and changes of properties in matter

149 A scientist heats a sample of metal to 640°C, and the metal turns to liquid. What could he conclude?

    a) 640°C is the metal's boiling point.

    b) 640°C is the metal's melting point.

    c) At 640°C the metal disintegrates.

    d) The metal has become a new compound.

Bonus: If the scientist continues to heat the metal, what will happen?

Properties and changes of properties in matter

150 You experiment with kitchen chemistry: You pour some water into a jar and add white vinegar, water, salt, and powdered sugar. You shake it up and the mixture looks the same as water. What type of mixture do you have?

    a) solvent   **c)** solution

    b) solute   **d)** dissolution

Properties and changes of properties in matter

151 A scientist tests one cup of liquid X and finds that it boils at 260°C. When she tests two cups of the same liquid, at which temperature would it boil?

    a) 130°C   **c)** 520°C

    b) 260°C   **d)** It's impossible to tell.

Properties and changes of properties in matter

152 This diagram shows the mass of the ice and its dish on the balance. When the ice melts, what will be the mass of the water and the dish?

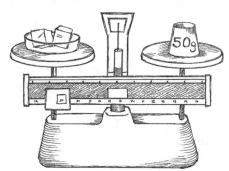

a) more than 100 g

b) more than 50 g

c) 50 g

d) less than 50 g

Properties and changes of properties in matter

153 Which of these is a chemical change?

a) ice cream melting

b) your breath forming steam in winter

c) a plant using sunlight to make food

d) sunlight speeding up the evaporation of a bowl of water

Properties and changes of properties in matter

154 You start with a 1 g sample of an element called gold. You heat the gold, cool it, and test it by running electricity through it. What are you left with?

a) less than 1 g of a new substance

b) less than 1 g of gold

c) 1 g of a new substance

d) 1 g of gold

Properties and changes of properties in matter

155 A scientist boils liquid until it turns to gas. How did she affect the density of the sample?

a) The density did not change.

c) The density increased.

b) The density decreased.

d) It is impossible to say.

156 The chart below shows how long it takes 10 g of metal X to melt at different temperatures.

Melting Time for 10 g of Metal X

Temperature (degrees Fahrenheit)	Melting Time (minutes/seconds)
660	3 min 12 sec
700	2 min 34 sec
740	1 min 52 sec

How long would it take to melt metal X at 695 degrees F?

a) 4 min 18 sec

b) 3 min 12 sec

c) 2 min 39 sec

d) 2 min 29 sec

157 This picture shows a chemical reaction. What is the mass of the resulting sample?

a) 40 g

b) 50 g

c) 100 g

d) more than 100 g

Physical Science

158 If the experimenter lets go of this marble, it will move because it's being acted upon by which force?

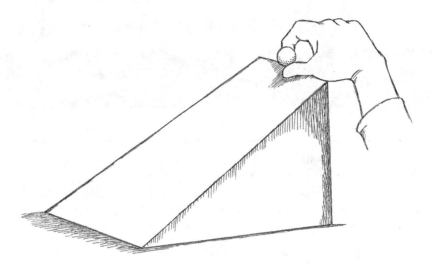

a) centrifugal force   c) gravity

b) momentum   d) friction

159 You conducted an experiment to see which object would travel the farthest after rolling down a ramp. Your results are shown here. Which object created the least amount of friction?

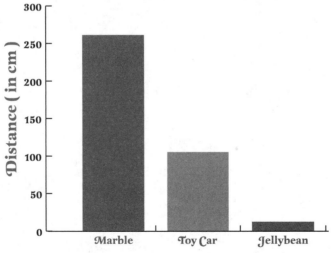

a) the marble   c) the jellybean

b) the toy car   d) all of the above

160 When you throw a ball, it follows the path shown here. Which force makes it change direction at the top of its flight?

a) momentum   **b)** inertia   **c)** air pressure   **d)** gravity

Bonus: Which path would the ball take if the above force did not exist?

161 Which diagram best shows what happens when a ball hits a wall? Circle your answer.

a)

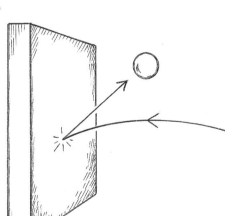

b)

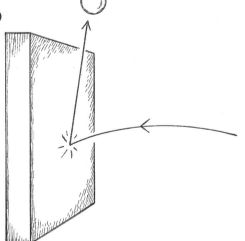

Bonus: Explain why you chose your answer.

Physical Science

162 In winter, ice and snow may cover the concrete sidewalk, making it slippery. It is harder to walk because ice lowers this force:

    a) gravity

    b) friction

    c) propulsion

    d) inertia

Bonus: Think about the force you chose above. What is one way to increase this force on an icy walk?

Physical Science

163 When you drop a ball, it bounces back up. What's the best explanation for why this occurs?

    a) The floor exerts velocity on the ball.

    b) The floor gives the ball momentum.

    c) The floor exerts a force on the ball.

    d) The floor takes the ball's energy.

Physical Science

164 How can a force affect a moving object?

    a) It can change an object's velocity only.

    b) It can change an object's velocity and direction.

    c) It can change an object's velocity and mass.

    d) It can change an object's velocity, direction, and mass.

Physical Science

165 A burning candle emits energy mainly in two forms. Name these two forms of energy.

166 In this scene, energy is being transferred from one form to another. Choose one of the instruments below. Explain in what form the energy started and in what form it became.

167 Which unit would best measure the stored energy that this jogger is using up?

a) calories

b) watts

c) joules

d) kilograms

Physical Science

168 Some chemical reactions produce heat. Where does the heat energy come from?

a) Energy is created.

b) Energy is taken from the atmosphere.

c) Kinetic energy is transformed into heat.

d) Chemical energy is transformed into heat.

169 Some power plants burn coal to provide the power you need in your home. What is one energy conversion that is taking place in the power plant?

a) fire to momentum

b) fire to gravitational energy

c) heat to thermal energy

d) heat to electrical energy

170 Which of these has energy?

a) a flying kite   **c)** a speeding train

b) a molecule   **d)** all of the above

171 Which diagram correctly shows the direction of heat flow?

a)

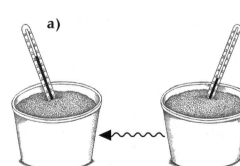

b)

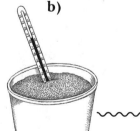

Physical Science

172 Which of these appliances costs the average family the most on its electric bill?

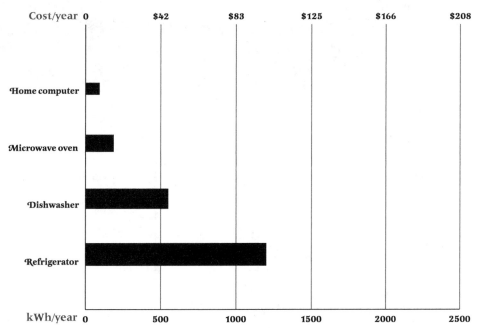

Physical Science

173 Burning coal is a chemical reaction called oxidation. Which of the following is true about this chemical reaction?

a) Energy is transferred into the system.

b) Energy is transferred out of the system.

c) Energy is not transferred into or out of the system.

d) No energy is involved.

Physical Science

174 When energy travels through a wire and lights a bulb, what is happening?

a) Energy is moving from place to place.

b) Energy is changing forms.

c) Energy is being lost as heat.

d) All of the above

175 This chart shows the forms of energy involved in running an electric egg beater. Fill in the chart to show the forms of energy involved in the other items.

	Heat	Light	Electrical energy	Kinetic energy	Potential energy due to gravity
an electric egg beater			X	X	
a burning candle					
a lit light bulb					
a falling object					
a pinwheel in the wind					

176 Name the energy source that starts as heat, travels as light, and provides the energy for all of Earth's food chains.

177 You want to know which material has a lower melting point—ice or ice cream. Which experiment can help you to investigate that?

    a) Put 1 g of ice and 2 g of ice cream in identical containers in a sunny spot and see which one melts first.

    b) Put 1 g of chocolate ice cream and 1 g of vanilla ice cream in identical containers in a sunny spot and see which   one melts first.

    c) Put 1 g of ice and 1 g of ice cream in the microwave oven for 3 minutes.

    d) None of the above

Bonus: Choose one experiment above that does not compare ice and ice cream melting points. Then make a correction so that it would work.

178 You set up an experiment to test how a red filter will affect the light from a lamp. You put a red piece of plastic in front of a light. Everything in the room turns red. If you wanted a control for this experiment, it would be

a) a different light with a red filter.

b) a different light with a green filter.

c) the same light with a green filter.

d) the same light with no filter.

179 You want to test paper-towel brands to see which is the most absorbent. You add 5 tablespoons of water to Brand A and find that it can absorb 2-$\frac{1}{2}$ tablespoons. How much water should you add to Brand B?

a) 2 $\frac{1}{2}$ tablespoons

b) 5 tablespoons

c) more than 5 tablespoons

d) $\frac{1}{2}$ cup

Bonus: Name one variable in this experiment.

180 Now, you test two properties of paper-towel brands. Based on these results, which brand would you buy?

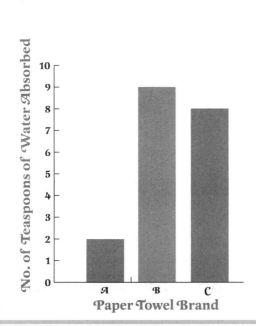
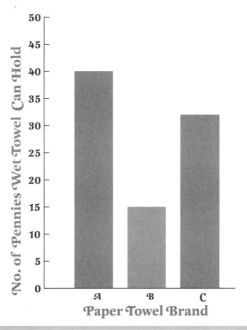

Answers

1. d

2. a

3. a

4. Plants can provide food, oxygen, shelter, or shade for animals. Animals can spread seeds, put nutrients into the soil, eat competing plants. Accept reasonable answers.

5. b

6. d; **Bonus:** When bees drink nectar they spread pollen from flower to flower—pollination happens before seeds form.

7. d

8. Most roots absorb water and nutrients and hold a plant in place.

9. c; **Bonus:** The one on the left is bone—bone is structured in concentric rings, making it light but strong. Muscle tissue, on the right, has long fibers that contract and relax.

10. b

11. d; **Bonus:** Photosynthesis

12. c

13. c

14. b

15. c

16. d

17. a

18. b

19. a

20. a

21. a

22. b

23. Examples of inherited traits include hair color, height, hair type, attached/unattached earlobes, skin color. Examples of learned traits include speaking certain languages, riding a bicycle, being a good cook. Accept reasonable answers.

24. b

25. Your heart will beat faster, you will breathe harder (both to get more oxygen to your muscles), your muscles will tense up. Accept reasonable answers.

26. a

27. d; **Bonus:** Defend itself by fighting, use protection such as porcupine quills or skunk's spray

28. Zookeepers could hide the food in a fake forest, give the gorillas lots of climbing equipment, etc. Accept reasonable answers.

29. c; **Bonus:** Eastern three-toed salamander

30. d

31. b; **Bonus:** Predators

32. d

33. Grass would increase if grasshoppers decrease; grass would decrease if grasshoppers increase, and shrews might increase.

34. Habitats provide food such as plants or animals that other animals need to eat; provide water for plants and animals; provide space to grow; and provide shelter.

35. Too many beavers could cut down

 Science Question of the Day Scholastic Teaching Resources

too many trees for a forest to remain healthy or dam a stream so much that it no longer flowed. Accept reasonable answers.

36. c

37. a; **Bonus:** Plants, trees

38. Anything that would help the creature conserve water (such as thick waxy skin or scales), gather water, or survive in sand. Accept reasonable answers.

39. Coloring helps hide a lizard from predators and possibly sneak up on prey; sense of smell and reflexes help it catch prey and run from predators; thick skin helps it conserve water.

40. d; **Bonus:** Accept reasonable answers.

41. a

42. c

43. c

44. Wash your hands frequently with soap and water, get enough sleep, eat well, don't get too close to people with colds. Accept reasonable answers.

45. c; **Bonus:** Sleep, exercise, avoiding too much stress

46. d; **Bonus:** Accept reasonable answers.

47. Helping people not get sick: 1, 3, 4 (primarily, though all are acceptable); helping people manage their weight: 4, 5, 6

48. d

49. d

50. d

51. d; **Bonus:** Tape measure or meterstick (Accept reasonable answers.)

52. a; **Bonus:** Accept reasonable answers.

53. b

54. b

55. b

56. b; **Bonus:** Accept reasonable answers. For example: If I try to approach both a chipmunk and a squirrel, then the squirrel will let me get closer before it runs away.

57. c; **Bonus:** Plants grow taller when they're not too crowded together.

58. b

59. c

60. d

61. Quartz

62. d; **Bonus:** Rocks wear away, dead plants and animals decompose. Accept reasonable answers.

63. b

64. Moonlight started at the sun, reflected off of the moon, and traveled to Earth.

65. c

66. c

67. d

68. d

69. c; **Bonus:** Look to the east.

70. b

71. b

72. d

73. b

74. c

75. c; **Bonus:** An egg has different layers that could represent the different layers inside Earth.

76. a

77. d

78. b

79. d; **Bonus:** All of the above.

80. c

81. b; **Bonus:** The atmosphere provides a moderate climate, protects the Earth from radiation, produces weather. Accept reasonable answers.

82. a

83. b; **Bonus:** The atmosphere traps the sun's heat.

84. Water; **Bonus:** Nearly 3/4 of Earth's surface is covered in water.

85. d; **Bonus:** Lava or melted rock can harden; dirt deep underground can be compressed by the great weight of earth and rock above it.

86. c

87. Water flowing or dripping (rivers, oceans, rain), wind (blowing sand fragments against rock), glaciers, friction (animals or people walking on it). Accept reasonable answers.

88. c

89. b

90. b

91. d

92. a

93. d

94. d; **Bonus:** In the air

95. a

96. a; **Bonus:** An earthquake, rain, a volcano erupting

97. b

98. c

99. c; **Bonus:** Some mountains would be bigger and some would be smaller.

100. a

101. The fossil at the bottom is the oldest.

102. a; **Bonus:** Accept reasonable answers.

103. c; **Bonus:** With our current technology, it could take hundreds of years.

104. c

105. a

106. d; **Bonus:** It creates tides.

107. c

108. Earth and Saturn are missing; Earth is the fifth bar with a diameter of 12,756 km; Saturn is the second bar with a diameter of 120,660 km.

109. d

110. Earth's orbit: 365 days; 1 moon; 24 hours in a day; diameter is about 12,800 km.

111. d

112. d

113. a; **Bonus:** Controlled variables include quantity of soil, type of container for soil, amount of water that's poured in. Variable that changes is the type of soil.

114. c

115. Soil A holds together best when wet. Soil C doesn't hold together at all. **Bonus:** Accept reasonable answers. Examples: You could pour a certain amount of water into a pot filled with each soil and see how much runs out the bottom to test how well each holds water. Or you could stick long twigs into each one, inserting them the same amount, to see how the different soils support twigs.

116. b; **Bonus:** Microscope and hand lens

117. All these tools could be used.

118. c

119. d

120. c

121. Color, shape, consistency (is it hard or soft? does it change shape easily?), mass, texture, smell (Accept reasonable answers.)

122. a

123. b

124. d; **Bonus:** You have to control all variables but one; for example, testing a ball of clay vs. boat-shaped clay. Accept reasonable answers.

125. Solid

126. a; **Bonus:** The liquid with the lowest density is on top.

127. b

128. d

129. a

130. Sample Y; **Bonus:** Mystery metal Z should be good at conducting electricity. Accept reasonable answers.

131. a

132. c

133. d

134. d

135. a

136. d

137. c

138. a

139. c

140. d

141. d; **Bonus:** Metal is a good conductor.

142. c

143. c

144. c

145. c

146. a; **Bonus:** Cold water cooled the air around it, so the air couldn't hold as much water. Water vapor from the air condensed on the glass.

147. d

148. Mass, color, and consistency stayed the same; shape and volume changed. Accept reasonable answers.

149. b; **Bonus:** With more heat, the metal will eventually boil.

150. c

151. b

152. c

153. c

154. d

155. b

156. c

157. c

158. c

159. a

160. d; **Bonus:** It will keep moving upward.

161. a; **Bonus:** When a ball hits a surface it bounces back at the same angle on the opposite side.

162. b; **Bonus:** Add something gritty such as dirt, sand, or salt. Accept reasonable answers.

163. c

164. b

165. Heat and light

166. Drum or guitar: kinetic energy to vibrations, or sound waves (Accept reasonable answers.)

167. a

168. d

169. d

170. d

171. b

172. Refrigerator

173. c

174. d

175.

	Heat	Light	Electrical energy	Kinetic energy	Potential energy due to gravity
an electric egg beater			X	X	
a burning candle	X	X			
a lit light bulb	X	X	X		
a falling object				X	X
a pinwheel in the wind				X	

176. The sun

177. d; **Bonus:** Accept reasonable answers. For example, choice "a" would work if you used 1 g of each substance (or if they had the same volume).

178. d

179. b; **Bonus:** The brand of paper towel and the amount of water it absorbed are variables.

180. Brand C